Globalization of S&T:
Key Challenges Facing DOD

Timothy Coffey and Steven Ramberg

Center for Technology and National Security Policy
National Defense University

FEBRUARY 2012

Timothy Coffey served as the Director of Research of the U.S. Naval Research Laboratory (NRL) from 1982 to 2001. From 2001 to 2007, he held a joint appointment as Senior Research Scientist at the University of Maryland and as the Edison Chair for Technology at NDU. He retired in 2001 and presently is under contract to NDU as a Distinguished Research Fellow.

Steven Ramberg is a Distinguished Research Fellow on assignment from the Applied Research Laboratory of Penn State University. At NDU, he occupies the Chief of Naval Research Chair. During his career, he served as a Fellow and as Vice President for Arete Associates from 2007 to 2010; as the Director of the NATO Undersea Research Centre in LaSpezia, Italy, from 2003 to 2007; and as Director and Chief Scientist for ONR from 2001 to 2003 after joining the Office of Naval Research in 1988. Before then, he worked at NRL, where he published more than 60 unclassified papers in the archival literature on fluid dynamics of bluff bodies, nonlinear ocean waves, stratified wakes, turbulence near a free surface, and related remote sensing topics.

TABLE OF CONTENTS

LIST OF FIGURES

LIST OF TABLES

Executive Summary

In the second half of the 20th century, the United States enjoyed stature and prosperity at levels seldom achieved in recorded history. The country's status included predominance in most fields of science and technology (S&T), as well as a phenomenal breadth and pace of innovation. We are now experiencing a global shift to a more level playing field among nations; demographics, economics, and political forces are the driving forces behind this shift. The impact of this shift on U.S. S&T will be significant. By the middle of the 21st century, it is likely that a number of nations will be similarly prosperous and technologically productive. No single nation or group will dominate as the United States did in the latter half of the 1900s. The U.S. share of the global S&T enterprise will decrease, and only a small fraction of U.S. scientists and engineers (S&E) will work on national security problems. This change poses challenges to the roles and conduct of Department of Defense (DOD) S&T. In particular, DOD's ability to maintain an authoritative awareness of S&T developments around the world will become increasingly problematic.

Most attempts to quantify these challenges utilize simple linear or exponential extrapolations. Although such approaches are helpful for short-term predictions, they tend to produce unrealistically pessimistic predictions for the timescales considered in this paper. The present work establishes an empirical relationship between an economy's gross domestic product (GDP) per capita and its ability to generate S&T knowledge. This paper then employs the results of a full economic analysis for the period 2005–2050 to estimate the S&T knowledge production for each of the world's 17 largest economies.

The estimate indicates that U.S. share of S&T productivity will decline from about 26 percent in 2005 to 18 percent in 2050. This decline, while problematic, is not unmanageable. At least through 2050, the United States will remain one of the world's most significant contributors to scientific knowledge. As a result, the U.S. S&T workforce should be large enough, relative to the world S&T workforce, to remain cognizant of S&T developments around the world—although the means of doing so may change. This ability to remain cognizant is important because by 2050, countries other than the United States will produce most scientific knowledge. Maintaining an authoritative awareness of S&T around the world will be essential if the United States is to remain economically and militarily competitive. This awareness includes the ability of the U.S. S&T workforce to authoritatively interpret trends in global S&T. For instance, the U.S. S&T workforce must be able to quickly recognize movements in the frontiers of knowledge and the potential for new military applications stemming from new knowledge or a combination of existing knowledge and new technology. The required awareness can be maintained only if the U.S. S&T workforce is a participant in the global S&T community. This is true for the DOD S&T workforce as well.

For DOD to succeed, it will be necessary to find a means to tap the knowledge of the larger U.S. S&T community regarding global S&T. It is only at this level that the United States will have a sufficient number of S&T "brain cells" to actually know what is occurring in the world of global S&T, what is important, and what is not important. Tapping this knowledge will be very challenging for DOD. Nevertheless, we must ensure the global S&T knowledge held by the larger U.S. S&T community is available to the military and that DOD has the internal capability to comprehend and exploit this knowledge through the DOD S&T workforce. The term "DOD S&T workforce" refers to those S&Es who are funded by DOD S&T dollars that fall into the categories of Basic Research (6.1) and Explorator y Development (6.2). This workforce is larger than the DOD federal S&T workforce (often calle d the in-house workforce), which, of course, has a special role. Som e members of the DOD S&T workforce will be em ployees of DOD; others will be involved through vehicles such as contracts, advisory committees, and cooperative programs with other government agencies.

Because national d efense is am ong the high est of the G overnment's responsibilities, it is essential that the S&T workfor ce supporting the national defense mission includes some of the nation's foremost S&Es. The people best able to maintain authoritative awareness of progress in S&T are those contributing to that progress. Such individuals m ust form the core of the DOD S&T workforce. A subset of this workforce shoul d also have an awareness of potential m ilitary applications. This awareness involves a worki ng knowledge of national security issues and general military concepts of operations (CONOPS) or concepts of use (CONUSE) in order to understand how something new *might* be useful or *could* become disruptive.

The largest U.S. window on the world of global S&T is its national S&T workforce. This workforce, however, spends m ost of its tim e not thinking about nation al defense. A properly designed DOD S&T workforce is the best condu it to the knowledge held by the U.S. national S&T community. The DOD S&T workfo rce should be sized and posit ioned to most effectively accomplish this task. In simple terms, the DOD S&T workforce m ust be p lugged into the national S&T community broadly (and to the exte nt possible into the global S&T community). To accomplish this, the DOD in-house S&T work force must be widely recogn ized for its contributions to the nation's S&T program . In other words, the DOD S&T workforce m ust be "card-carrying" members of the larger S&T comm unity. This recognition is paramount because the total number of DOD in-house S&T researchers will not increa se in proportion to the growth of the global S&T enterprise. Federal employment, compensation, and facility initiatives may be necessary to attract and retain the necessary talents to achieve thes e ends. DOD should establish organizational imperatives to prioritize a nd support authoritative global S&T awareness activities.

External researchers from academ ia, other government agencies, and in dustry must complement the DOD in-house S&T workforce. These individuals must also be card-carrying members of the global S&T community. They m ust be selected and funded by federal program managers who themselves are (or were) card-carrying members of the S&T community and who possess a keen awareness of the needs of the m ilitary. An effo rt should be m ade to establish a bon d between external researchers and DOD to extend the reac h and effectiveness of the DOD S&T awareness network. Indeed, while external researchers are likely to be f unded by other sources as well, a bond with DOD will yield better awareness functions for DOD. These bonds become a means for attracting highly qualified individuals to DOD employment. Furthermore, these bonds can enable DOD to better mobilize sectors of the larger S&T community in times of acute and pressing need and have a knowledge base suitable to preparation for the war after next.

To make the m ost effective use of the DOD S&T workforce, it will be necessary to em ploy emerging tools for tech nology forecasting (TF) a nd data mining. These tools should be m ade available at the desktop of a ll DOD S&Es engaged in the aw areness function. The tools should access the widest possible databases of technical reports across the United States and, with suitable security and IP protec tions, should include inform ation from all research proposals submitted to all U.S. agencies.

To accomplish the objectives outlined above, the competition between the need to control S&T information and the need for open S&T communicat ions must be m anaged in the transition to 2050 and beyond. Policies and procedures for info rmation control should be reevaluated to determine a strategic balance between the risk s, costs, and benefits of S&T inform ation control in a 2050 context.

Even if the DOD S&T community is reinvigorated as suggested above, the problems confronting DOD as a result of S&T globalization will be formidable. DOD will not have the fiscal resources to buy its way out of these problem s by funding its own standalone program that is large enough to maintain insight into the global S&T program or to play "catch up" to a foreign effort that has gotten ahead. Some nonmonetary means must be found to motivate the national S&T community to accept som e responsibility for keeping DOD aw are of global S&T developm ents that could have significant national defense implications. In this regard, most of the U.S. nationa l S&T workforce of 2050 is yet to be educated. Perhap s the education system is where the motivation should be developed as part of the effort to imbue students with an understanding of their profession's civic responsibilities. A concern for the health of national defense should be am ong those civic responsibilities. It m ay be that an acceptance of this civic responsibility among the national S&T co mmunity is essential for the solution of the DOD g lobal S&T awareness problem.

1. INTRODUCTION

In the second half of the 20th century, the United States enjoyed stature and prosperity at levels seldom achieved in recorded history. The country 's status included predominance in most fields of science and technology (S&T), as well as a phenomenal breadth and pace of innovation. We are now experiencing a global shift to a more level playing field among nations; demographics, economics, and political forces are the driving forces behind this shift. The impact of this shift on U.S. S&T will be significant.

The globalization of S&T has been a topic of discussion for some time. Some of the discussion has related to its impact on U.S. competitiveness and often argues to increase the supply of scientists and engineers (S &E) and increase funding for the scientific enterprise. [1] Other discussion has focused on how globalization is changing how S&T occurs and often addresses the need to prepare scientists, engineers, and societies for these changes. [2]

By the middle of the 21st century, it is likely that a number of nations will be similarly prosperous and as technologically productive as the United States, and no single nation or group will dominate as the United States did in the latter half of the 1900s. During the period of U.S. dominance, nearly half of the S&Es performing research in the world were in the United States. Furthermore, a substantial fraction of these S&Es were working on research for national security and funded by the Department of Defense (DOD). By 2050 this situation will have changed dramatically. The U.S. share of the global S&T enterprise will have decreased, and only a small fraction of U.S. S&Es will work on national security problems. This change poses challenges to the roles and conduct of DOD S&T. Before addressing these challenges, it is helpful to attempt to quantify the magnitude of the shift underway and the likely timeframe over which it will occur.

---

[1] *Rising Above the Gathering Storm: Energizing and Employing America for a Brighter Economic Future*, National Research Council (NRC) Report (2007); see http://www.nap.edu/openbook.php?record_id=11463&page=1.
[2] *Knowledge, Networks, and Nations: Global Scientific Collaboration in the 21st Century*, The Royal Society (2011); see http://royalsociety.org/uploadedFiles/Royal_Society_Content/Influencing_Policy/Reports/2011-03-28-Knowledge-networks-nations.pdf.

2.   THE CHANGING CONTEXT FOR U.S. AND DOD S&T

Economically advanced and developing countries   have widely recognized that S&T is a key driver in the world economy and in military affairs. Most countries now have stated objectives to invest 2 percent to 4 percen   t of their gross dom   estic product (GDP) into research and development (R&D).[3] For decad es the United States   has been the predom inant force in th e application of S&T to econom ic and m ilitary advancement through an averag e investment of about 2.5 percent of GDP into R&D. At present,   the United States has the largest GDP of any single county and accounts for about 23 percent of the world GDP. [4] However, the U.S. fraction of world GDP is expected to decline over the com   ing decades.[5] Since it is unlik ely that th e United States will markedly increase its R&D investm ent as a percen tage of its GDP, it f ollows that U.S. investment in R&D will declin e as a percentage of the world investm ent in R&D. This anticipated decline has led to m uch speculation about the Un ited States' ability to m aintain a dominant position in S&T.[6, 7] It would be helpf ul if one co uld estimate the magnitude of this expected decline for the coming decades.

The following section provides a rough estim ate of the likely evolution of the U.S. position in S&T through 2050. W e consider S&T to be the subs et of R&D that deals with basic research (e.g., DOD funding category 6.1) a nd exploratory develop ment (e.g., DOD fund ing category 6.2). This is often referred to as the "Tech Base," but in this paper it will be referred to as S&T. There is no "first principles" m ethodology available to make the desired estim ate of the future U.S. position. However, it seems reasonable to assume that an economy's ability to reach its full potential for the generation of S&T knowledge is related to that economy's standard of living. An economy with a low standard of living is unabl e to support the infrastructure needed to reach its full potential. This situation will change as the economy becomes more advanced. When the standard of living becom es typical of an a dvanced economy, then that econom y's ability to generate S&T knowledge should become typical of advanced economies. In this regard, the most often used m etric for a n economy's standard of living is the GDP per capita. Although this metric has many limitations, it does provide a means by which to compare standards of living in an average sense. That comparison should be adequate for the predictions needed in this paper.

Because the United States has th e largest economy of the econom ically advanced countries, we will consider it to be ty pical of an advanced economy and will, th erefore, provide comparisons relative to the United States. In this regard, it should be noted that fina ncial comparisons among countries are often perform ed using m arket exchange rate (MER) comp arisons or purchasing power parity (PPP) comparisons. [8] The comparison method chosen can lead to different results. However, the two rates are usually quite sim ilar among economically advanced countries. As an example of GDP per ca pita comparisons among advanced economies and emerging economies,

---

[3] *OECD Science, Technology, and Industry Outlook 2010;* see http://www.keepeek.com/Digital-Asset-Management/oecd/science-and-technology/oecd-science-technology-and-industry-outlook-2010_sti_outlook-2010-en.

[4] *CIA Fact Book;* see https://www.cia.gov/library/publications/the-world-factbook/.

[5] *ERS/USDA Data: International Macroeconomic Data Set;* see http://www.ers.usda.gov/Data/Macroeconomics/.

[6] *Rising Above the Gathering Storm: Energizing and Employing America for a Brighter Economic Future* (2007).

[7] *Knowledge, Networks, and Nations: Global Scientific Collaboration in the 21st Century,* The Royal Society (2011).

[8] For a discussion of MER and PPP, see T. Callen's "PPP Versus the Market: Which Weight Matters"; see http://www.imf.org/external/pubs/ft/fandd/2007/03/basics.htm.

Figure 1 provides a historical ME R comparison of GDP per capita among the United S tates, Organization for Economic Cooperation and Development (OECD) nations, Japan, and China. The data is normalized to U.S. GDP per capita.

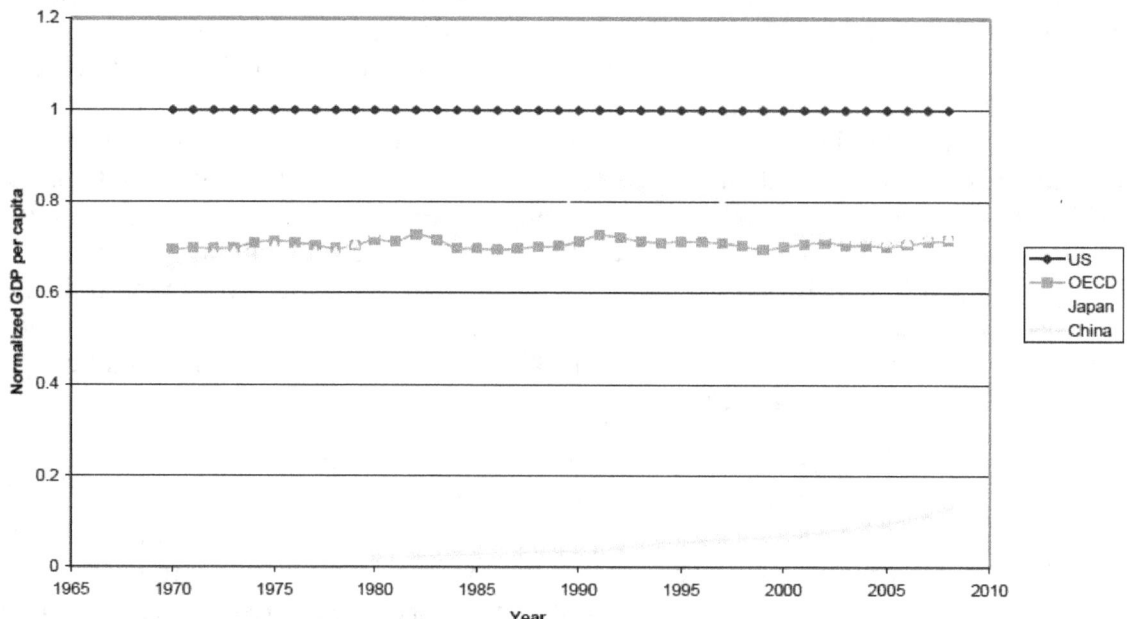

Figure 1. GDP per Capita for OECD, Japan, and China (normalized to United States)[9]

As shown in Figure 1, although there is som e GDP per capita spread am ong the advanced economies, they are clustered near the Unite d States. The em erging economy (China) shows GDP per capita that is markedly less than that which is typical of advanced economies. Based on the above arguments, we expect that a develo ping economy should show a lower per capita rate of scientific knowledge generation than the advanced economies.

It would be helpful to know whether or not a correlation between the rate of scientific knowledge generation and standard of living as measured by GDP per capita actually exists. Although there is no definitive m easure of S&T knowledge ge neration, it is generally accepted that S&T publications and patents are i ndicative of the gene ration of S&T knowledge. We therefore postulate that the gene ration of scientific publications and patents is a surrog ate for the generation of S&T knowledge. To gain some in sight into the relationship betw een GDP per capita and research ou tput, Figure 2 displays the res earch publication output per m illion inhabitants for Japan, China, and the total of the OECD countries measured relative to the United States for 2008 (the actual U.S. number for 20 08 was about 900 research publication per m illion inhabitants). Also included in Figure 2 is the GDP per capita calcula ted by the PPP m ethod and the MER method (both normalized to the United States).

---

[9] *OECD Fact Book 2010.*

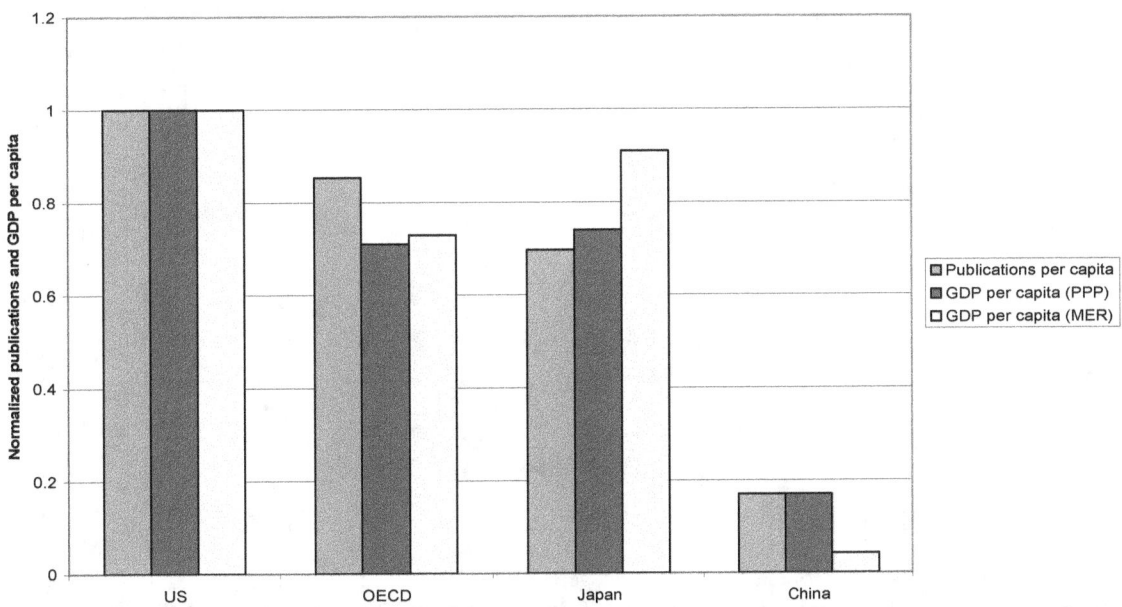

Figure 2. Comparison of Publications and GDP per Capita (normalized to United States)[10]

Figure 2 suggests that the normalized publication output correlates with the normalized GDP per capita and that the PPP m ethod of calculating the GDP per capita provides a slightly better correlation than does the MER m ethod. The latter po int may be peculiar to the year chosen for comparison, or it may indicate that the PPP m ethod provides a better measure of the standard of living of an economy and that the standard of living underlies an economy's ability to sustain the infrastructure needed to generate S&T knowledge.

Regarding patent productivity, Figure 3 displays trilateral patents [11] per m illion population for several economically advanced co untries and the to tal for OECD countries norm alized to th e United States for 2007. China is not a significant pl ayer in the trilateral patent arena but is included for comparison.

---

[10] *OECD Science, Technology, and Industry Outlook 2010.*
[11] Trilateral Patent Offices; see http://www.trilateral.net/about html.

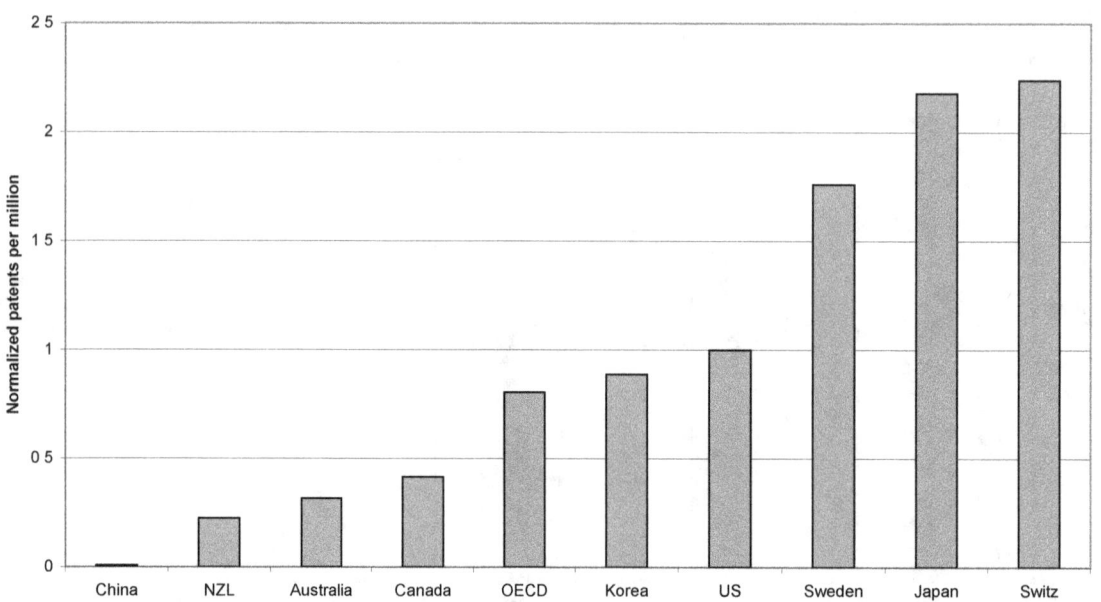

Figure 3. 2007 Triplaterial Patents per Million Population for Several Economically Advanced Economies (normalized to U.S. patents per million of 52.6)[12]

Figure 3 suggests that, within a factor of two, economically advanced countries produce similar numbers of patents per inhabitant and emerging economies produce substantially fewer patents per inhabitant. Unfortunately, em erging economies are not yet sign ificant players in the generation of trilateral patents. Hence, a meaningful comparison between patent productivity and GDP per capita is not available for thos e economies. It is e xpected, however, that em erging economies, as they becom e more advanced, will te nd toward patent productivity sim ilar to that of economically advanced economies. Here again, the use of GDP per ca pita normalized to the GDP per capita of econom ically advanced economies should provide a param eter that indicates an economy's ability to produce S&T knowledge.

If one accepts that research publications and p atents are a reasonab le surrogate for scien tific knowledge generation, then Figures 2 and 3 suggest that the rate of change of S&T knowledge can be expressed by the following simple differential equation:

$$\frac{dK}{dt} = \eta P, \ (1)$$

where K represents S&T knowledge, P represents an economy 's total population, and η is an efficiency factor for the gene ration of S&T knowledge. The efficiency f or advanced economies is unlikely to vary by more than a factor of two.

If we focus on research publica tions and assume the econom y's S&T efficiency factor scales linearly with GDP per capita, then:

$$\eta = 900 \frac{GDP / capita}{USGDP / capita} \ (2)$$

---

[12] *OECD Science, Technology, and Industry Outlook 2010.*

The quantity 900 represents a number typical of U.S. research publications per year per million inhabitants. According to equations 1 and 2, knowledge of an economy's GDP per capita relative to U.S. GDP per capita allows one to make a rough estimate of that economy's ability to generate scientific knowledge. The viability of this approach is, of course, based on the empirical relationship established by Figure 2.

To use equation 2, one must predict the GDP per capita. Simple linear extrapolations or past growth rates are often used to make such predictions. Unfortunately, these approaches are helpful only for short-term predictions but are not valid for the extended period considered here. Predicting GDP per capita for extended periods is a complex undertaking and is beyond the scope of this study. A proper prediction must include, among other things, expected exchange rate dynamics, long-term demographic trends, education trends, and investment rates. A recent study by Hawksworth and Cookson made such predictions for the period 2005–2050.[13] This study projected the size of the 17 largest economies in the world. The study included 10 advanced economies (United States, United Kingdom, Japan, Germany, France, Italy, Canada, Spain, Australia, and South Korea) and the 7 largest emerging economies (China, India, Brazil, Russia, Indonesia, Mexico, and Turkey). The analysis was performed using both market exchange rates and purchasing power parity. We will use the data presented by Hawksworth and Cookson utilizing PPP rates. Using market exchange rates, although providing quantitatively different results, does not change the basic conclusions. This prediction was published in 2006, yet we expect the relative sizes forecast for the various economies will likely be preserved even though the timeframe may shift slightly as a result of the "Great Recession" that began in late 2007. As long-term economic analyses that reflect the recession become available, it will be straightforward to adjust the prediction made here.

Figure 4 presents the results for the predicted GDP per capita normalized to the predicted U.S. GDP per capita and using PPP rates over the period 2005–2050.

---

[13] *The World in 2050: Beyond the BRICs—A Broader Look at Emerging Market Growth Prospects*, Hawksworth and Cookson, PricewaterhouseCoopers (2006); see http://www.pwc.ch/user_content/editor/files/publ_tls/pwc_the_world_in_2050_e.pdf.

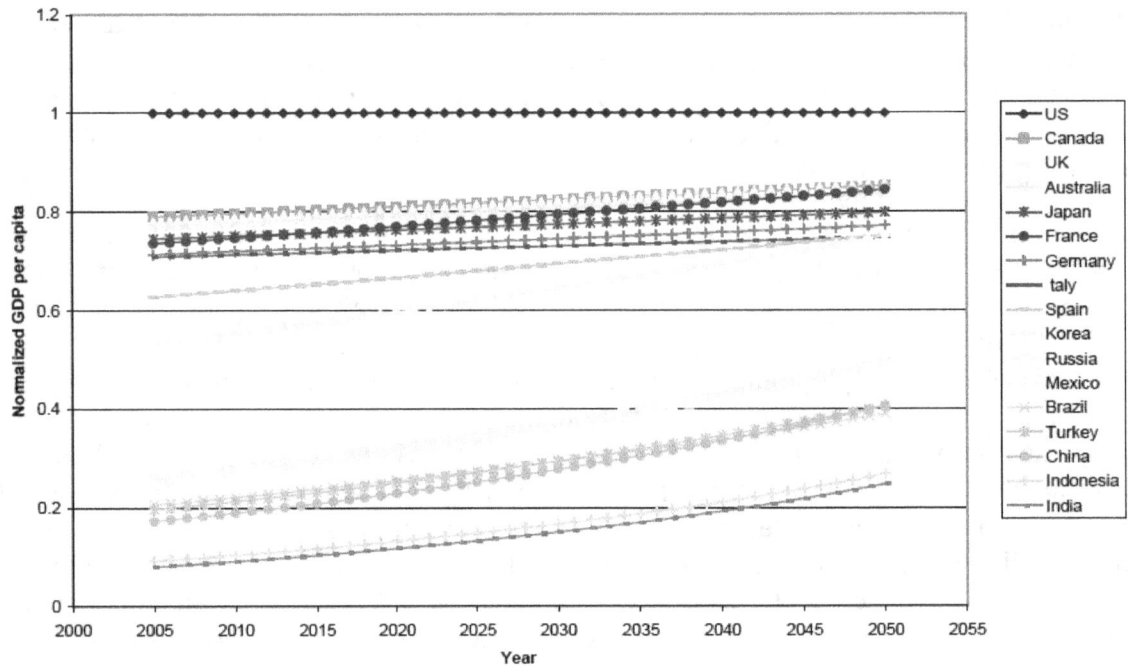

Figure 4. GDP Per Capita (normalized to United States using PPP rates)[14]

The data presented here result from a simple exponential fit to the data presented by Hawksworth and Cookson for 2005 and 2050. The clear separa   tion between the econom ically advanced economies and the emerging econom ies is evident. It is also evident that, although the gap between the advanced and e merging economies narrows significantly, no e merging economy actually enters the economically advanced category by 2050.

The data presented in Figure 4 and equation 2   allow one to estim ate the annual num ber of research publications by each of the 17 econom ies studied by Ha wksworth and Cookson for the period 2005–2050. From these estim ates, one can cal culate the fraction of the total num ber of research publications contributed by each of the 17 economies per year. Figure 5 presents this information. The fraction is com puted relative to the total num ber of research publications predicted for all 17 economies considered.

---

[14] Ibid.

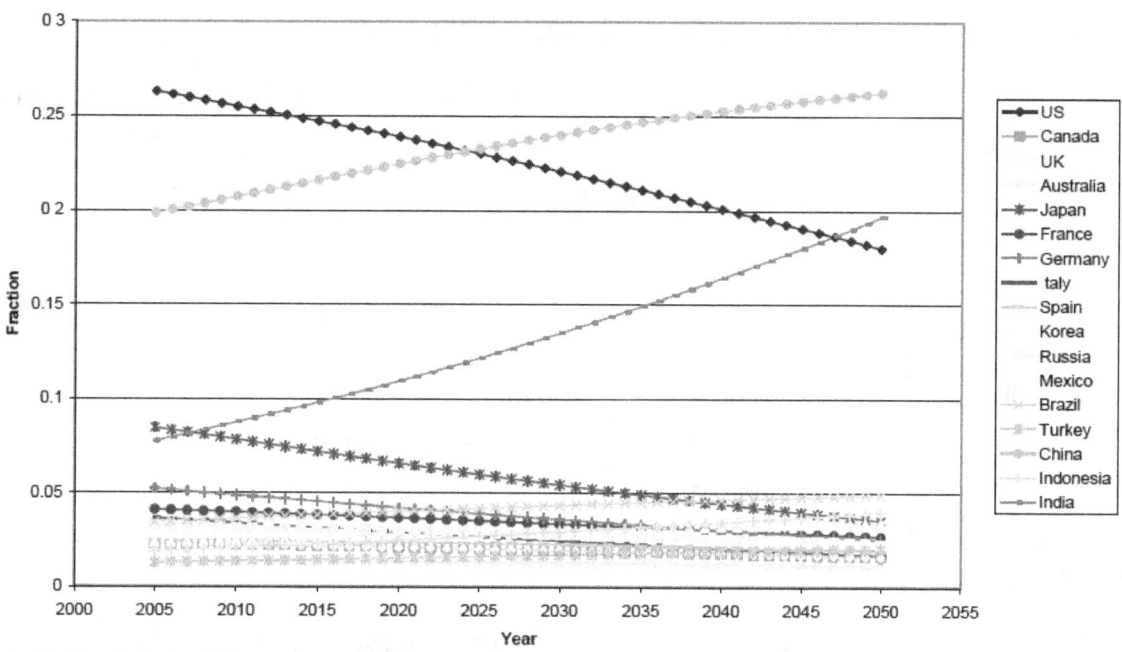

Figure 5. Estimated Fraction of World Research Publications for 17 Economies Over the Period 2005–2050

According to the predictions shown in Figure 5, by 2050 three economies (China, India, and the United States) will account for about two-thirds of the world's research output.

As noted above, Figure 5 results from an empirical relationship between an economy's GDP per capita and its scientific productivity and is based on a particular economic analysis. Its utility derives from the trends and timescales that it identifies rather than from the details for any particular year. It is the trends and timescales that are helpful in quantifying the changes facing the United States. Figure 5 suggests that the U.S. share of world S&T knowledge production will decline from about 26 percent in 2005 to 18 percent in 2050. This decline, while problematic, is not unmanageable. At least through 2050, the United States will remain one of the world's most significant contributors to scientific knowledge. As a result, the U.S. S&T workforce should be large enough, relative to the world S&T workforce, to remain cognizant of S&T developments around the world. This ability to remain cognizant is important because, by 2050, countries other than the United States will produce most scientific knowledge. Maintaining an authoritative awareness of S&T around the world will be essential if the United States is to remain economically and militarily competitive. This awareness is possible only if the U.S. S&T workforce is a participant in the global S&T community. This is true for the DOD S&T workforce as well. In this regard, the DOD workforce will face several issues. Before getting into these issues, it is helpful to review the purpose of and priorities for DOD S&T.

3.   PURPOSES AND PRIORITIES OF DOD S&T

To frame the challenges facing DOD S&T as a result of globalization, it is first helpful to outline the purposes for which DOD invests in S&T:

- Maintain awareness of the broad global S&T program.
- Prepare for the war after next.
- Invest in S&T exploration.
- Discover things.
- Invent things.
- Position DOD to be able to mobilize the larger S&T community, if required.
- Find out early what will not work.
- Prepare for the next war (e.g., support acquisition of new systems).
- Support the needs of existing systems.

These items are not n ecessarily the custom ary or official reasons for D OD S&T, nor are they entirely independent of one another. They ar e, however, often used to describe the DOD S&T portfolio. Most of these activities and concepts unde rlie the sem inal report to P resident Truman from Vannevar Bush in 1946, which fra med the highly successful U.S. S&T strategy for nearly 50 years.[15] That strategy can be summarized as follows: spend as m uch as you can gainfully invest in a broad rang e of topics so you don't com e in second place when it cou nts. Overt military relevance was a secondary considera tion and, in fact, was viewed as a possible impediment to truly basic research endeavors. This s trategy was m otivated by experience in World War II (WWII), where opponents held several technical forefronts (e.g., precision-guided weapons, jet engines, rocket propulsion, armor) throughout the conflict. It was only through the superior numbers and the industrial m ight of the Allies that WWII ended on term s favorable to the Allies. This strategy also served the Nation well throughout the Cold W ar. However, it should be noted that evidence suggests this strategy may be unsustainable in the 21st century and that relevance or societal m otivation may be as important as the character of the research and should be a factor in strategy formulation.[16]

The above list is presented in priority order, in the authors' opinion, for the DOD S&T portfolio and reflects the importance of the research and its relevance. We arrived at this prio ritization in two ways. F irst, the upper items are larg ely in the competency of S&T with little or no o ther providers in DOD; the lower item s are increasingly found in the core competencies of other DOD functions. Second, the ability to fully addr ess the lower item s often depends on one or more skills and elem ents of knowledge found in th e activities of the upper item s. The following bullets expand on these purposes and introduce th e likely challenges to DOD in a chieving the various purposes:

- ***Maintain awareness of the broad global S&T program.*** This awareness includes the ability of the U.S. S&T workforce to authoritatively interpret trends in global S&T. For

---

[15] *Science the Endless Frontier,* Vannevar Bush (1945); see http://www.nsf.gov/od/lpa/nsf50/vbush1945.htm.
[16] See, for example, *Pasteur's Quadrant: Basic Science and Technological Innovation,* D.E. Stokes, Brookings Institution Press (1997).

instance, the U.S. S&T workforce must be able to quickly recognize movements in the frontiers of knowledge and the potential for new military applications stemming from new knowledge or a combination of existing knowledge and new technology (i.e., avoid surprises and exploit quickly). A recent publication characterized S&T as involving two major phases: prospecting and mining.[17] Prospecting can be thought of as an activity that is important in the long term but shows little or no return on investment in the short term. Mining begins when prospecting has progressed to the point where sufficient understanding has been gained such that product (or system) development programs can begin using traditional return on investment metrics.[18] To maintain awareness, the U.S. S&T workforce must quickly spot the movement from one phase to another; this is possible only through involvement in the appropriate prospecting phases. Spotting this movement was relatively easy for the United States and DOD 30 years ago, but it may be more challenging now as the largest share of all prospecting advances occurs outside of the United States. A simple example of a new challenge is that much of the work may be reported in languages other than English.

- ***Prepare for the war after next.*** In many ways, this is where DOD S&T has a key role to play. The activity here is well beyond the horizon of current military requirements, and the timescales involved are typical of S&T (e.g., 10–20 years) that includes a prospecting phase. This activity involves developing future capabilities scenarios, alternatives, and likelihoods based on estimates of technology trends and new knowledge gains foreseen in the research and exploitation pipelines. It plays an important part in determining areas for S&T investments, motivating discovery and invention, and involving the larger S&T community so it is prepared to contribute to the required efforts when needed. Playing in this area requires highly talented and respected S&Es. A growing challenge here is attracting such individuals to participate when most S&T is being done outside of DOD.

- ***Invest in S&T exploration.*** The U.S. Government, including DOD, has a strong record of investing in research at the frontiers of S&T (prospecting) to foster the discovery of new knowledge. DOD is unusual with regard to federal funding of S&T efforts in that it may also be the user of the results of its S&T efforts. Numerous examples of this scenario are found in military systems (e.g., radar, sonar, missiles, aircraft). One cannot expect other sources (e.g., industry) to invest significantly here, and studies indicate this is increasingly the purview of governments as industry looks to early returns on its R&D investments.[19] Many use the term "curiosity-driven" to describe this activity, but "knowledge-driven" seems more appropriate to the actual motivations of both the researchers and the sponsors. In an era of much broader global S&T, the U.S. share of exploration will be less. This presents the nation and especially DOD with a dilemma: the reduced share suggests a narrowing of topics in which S&T investments are made, but the mission requires that awareness cover all areas of the global S&T program. Resolution of this dilemma may require new approaches for investing in S&T.

- ***Discover things.*** An S&T portfolio ranging from exploration to application-driven investments is barren without frequent successes in the nature of discovery. Successes

---

[17] *The S&T Innovation Conundrum,* T. Coffey, J. Dahlburg, and E. Zimet, National Defense University (NDU) Defense & Technology Paper Number 17 (2005).
[18] *Avoiding Technology Surprise for Tomorrow's Warfighter:* A Symposium Report, NRC (2009).
[19] *OECD Science, Technology, and Industry Outlook 2010.*

can be large or incremental advances and may not fully coincide with the original expectation of the research. These are the characteristics of the S&T enterprise and should be acknowledged and fostered explicitly. S&T prospecting activities that lead to mining coincide with a discovery in some cases, but not always. Discovery success rates for a given portfolio size will be influenced by the quality, efficiency, and connectivity of the S&T workforce, but otherwise the nature of discovery will not change in a more global S&T enterprise.

- *Invent things.* One can consider a discovery with clear and compelling application(s) as an invention. However, invention does not necessarily involve a discovery. It may, for example, involve the realization that a certain combination of well-known technologies provides a novel capability. Much of the material in the previous bullet is equally valid here, and, again, the basic nature of invention will not change in the coming decades.

- *Position DOD to be able to mobilize the larger S&T community.* The mobilization of the S&T community was complete in WWII, and the desire to retain that ability was an aim of Vannevar Bush's report to Truman.[20] During the Cold War, large S&T investments preconditioned the U.S. ability to mobilize segments of the research community when new capabilities were sought (e.g., the space race) or new threats emerged (e.g., strategic deterrence). One could argue that an attempt to mobilize a very willing research community post-9/11 was not very successful because limited institutional preconditioning existed for a response to terrorism. Such broad preconditioning for national security will be even more problematic in 30 years when DOD is a smaller fraction of the U.S. S&T investment and the United States is one of a half-dozen similar players globally. It seems a new approach must be developed to accomplish this purpose. One consideration could be to develop more efficient and rapid means of mobilization involving coordination across U.S. agencies engaged with the non-DOD S&T community. Such a means for broad mobilization would be useful outside the DOD mission area as well (e.g., homeland security, environmental crises).

- *Find out early what will not work.* This important S&T purpose is often overlooked, risking large system engineering overruns and procurement delays. An important role of S&T is providing a sanity check on the promises and claims of new system concepts, as well as the maturity of component technologies for incorporation into a system, to enable DOD to be a "smart buyer." This role demands authoritative, trusted, and objective S&T expertise at a very early stage in areas where new concepts are being proposed. Trusted agents should not have vested interests (pro or con) regarding the new concepts. Such a role is unchanged when DOD is a smaller player in the global S&T enterprise. A problem that arises, however, is sufficient access to a cadre of trusted S&Es who have the required technical authority and breadth but do not have some conflict of interest.

- *Prepare for the next war.* This purpose primarily involves acquiring new military systems for delivery in 5 or more years against known, defined requirements. Early on, the S&T role is largely proof-of-concept and prototyping in relevant environments. Thereafter, the component technologies are—or should be—well known, and the S&T role involves mostly minor ("spiral") upgrades and risk reduction activities. In terms of Technology Readiness Levels (TRL), this purpose is in the range of TRL 4 to 6 and

---

[20] *Science the Endless Frontier,* Vannevar Bush (1945).

generally beyond the S&T portfolio. These activities should scale to the magnitude of system procurement activities provided sufficient U.S. expertise is available.

- *Support the needs of existing systems.* Systematic upgrades to existing major systems, as well as quick fixes and one-off supplements to large capabilities, can be a role of mature S&T. One can argue that this activity has grown in recent years because existing systems could not accurately plan for the war after next. That said, the functions here are unlikely to differ much from today versus 2030 and beyond.

In summary, Table 1 assesses our nine reasons for DOD S&T investments. We indicate by red, yellow, or green our assessment of the challenge(s) to DOD in achieving that purpose in 2030 and beyond. Green means the current practices are likely to suffice, and red indicates that challenges to these purposes are significant. Yellow is used to highlight purposes that could be constrained by relevant S&T expertise available to DOD or related challenges. Further, we indicate whether DOD S&T must take a lead for each purpose or whether it can rely on the lead of others in DOD or elsewhere.

Table 1. Priorities for DOD S&T

Topic	Most Important (i.e., DOD S&T must lead for itself)	Important (i.e., DOD S&T must participate but can rely on others to lead)
Maintain awareness of the broad global S&T program	X	
Prepare for the war after next	X	
Invest in S&T exploration		X
Discover things		X
Invent things		X
Position DOD to be able to mobilize the larger S&T community, if required	X	
Find out early what will not work	X	
Prepare for the next war		X
Support the needs of existing systems		X

DOD's ability to maintain authoritative S&T awareness is key to sustaining success overall. This ability faces great challenges in 2030 and beyond. Only DOD can maintain S&T awareness for its own aims. Therefore, we focus on this subject in the remainder of this report.

4. MANAGING THE TRANSITION TO 2030 AND BEYOND

The United States has been the dominant force in S&T since WWII. It is clear, however, that this dominance is being replaced by a situation that is better described as a shared participation in the generation of S&T knowledge. In the near term, as a result of its large inv estment in S&T infrastructure and the high im pact of U.S.-p roduced S&T knowledge, the United States will maintain its leading pos ition in spite of the rapid growth of global S&T. We will maintain this position largely because it takes d ecades to p ut in place a S&T engine com parable to that currently available to the United States. In the long term, however, U.S. dominance must decline. This is s imply a matter of numbers. The em erging economies will grow much faster than the United States' economy.

As the emerging economies become more advanced, they will be inc reasingly able to invest in S&T and have established goals to do so.[21] They will learn from U.S. experience and improve on it. If the world rem ains stable (e.g., no m ajor wars, no cataclysm ic natural disasters), over the next 50 years the largest em erging economies will surpass the U.S. economy in size and will b e positioned to inves t larger sums in S&T th an the United States. This realization has caus ed concern in many quarters. It is difficult to accept that the United States will not b e the dominant force in S&T for m uch longer since we have never known anything else. It is im portant, however, that the United States develop a realis tic and pragm atic approach to this transition. Doing so will be essential to obtaining an aut horitative S&T understanding to properly evaluate developments so the Nation can remain militarily and economically competitive as technology evolves. Failure to do so will like ly have sig nificant negative outcomes for U.S. econom ic competitiveness and national security. Perhaps th e first thing to understand is that the United States is in no imminent danger of being eclipsed in S&T and it can remain a major force in S&T through the 21st cen tury. For this to happen, t hough, the United States will n eed to ach ieve balance among legitim ate but often com peting demands. Some early indicati ons in this regard are troubling. This section discusses som e of the challenges and com peting demands and offers suggestions on how to proceed while we are in transition to the new global S&T mix.

Although the situation f or the Unit ed States at large seem s manageable, the situation for DOD seems more problematic. Coffey has discussed the reason for this,[22] which derives from the fact that the DOD in-house S&T workforce has declined much more rapidly than has the U.S. S&T workforce relative to the global S&T workforce. This decline is likely to continue at some level even if current trends improve.

The issue confronting DOD is that of putting in place the necessary S&T workforce and properly empowering that workf orce. This task can be da unting to even visualiz e. We offer a s imple visualization in Figure 6 in which the area s of the segm ents are rep resentative of size. The numbers shown are our simple estimates and are intended only to give some sense of the scale of the problem. It is helpful to consider the problem as a series of layers. Layer 1, or bottom layer, is the global S&T community over which awar eness is needed. Layer 2 is the U.S. S& T community, which, collectively, should have a good understanding of what is happening in the global community. Layer 3 consists of the nongovernment S&T community that is bonded

---

[21] *S&T Strategies of Six Countries—Implications for the United States,* NRC Report (2010).

[22] *Building the S&E Workforce for 2040: Challenges Facing the Department of Defense,* T. Coffey, NDU Defense & Technology Paper Number 49 (2008).

(through contract, past association, etc.) in some way with DOD. Layer 4 consists of the in-house S&Es who are DOD employees funded by S&T dolla rs. Strictly speaking, Layers 3 and 4 are also members of Layer 2, which itself is a   member of Layer 1. The nu mber of in-house DOD S&Ts and the nongovernm ent S&Ts bonded to   DOD generally track the size of the DOD program. Because the DOD program, on average, has been constant for decades, we assume this will continue. We have, theref ore, kept the siz e of Layers 3 and 4 at the presen t values. One could add m any additional com ponents (e.g., S&Es of Al lies), but for conceptual purposes the four layers should suffice.

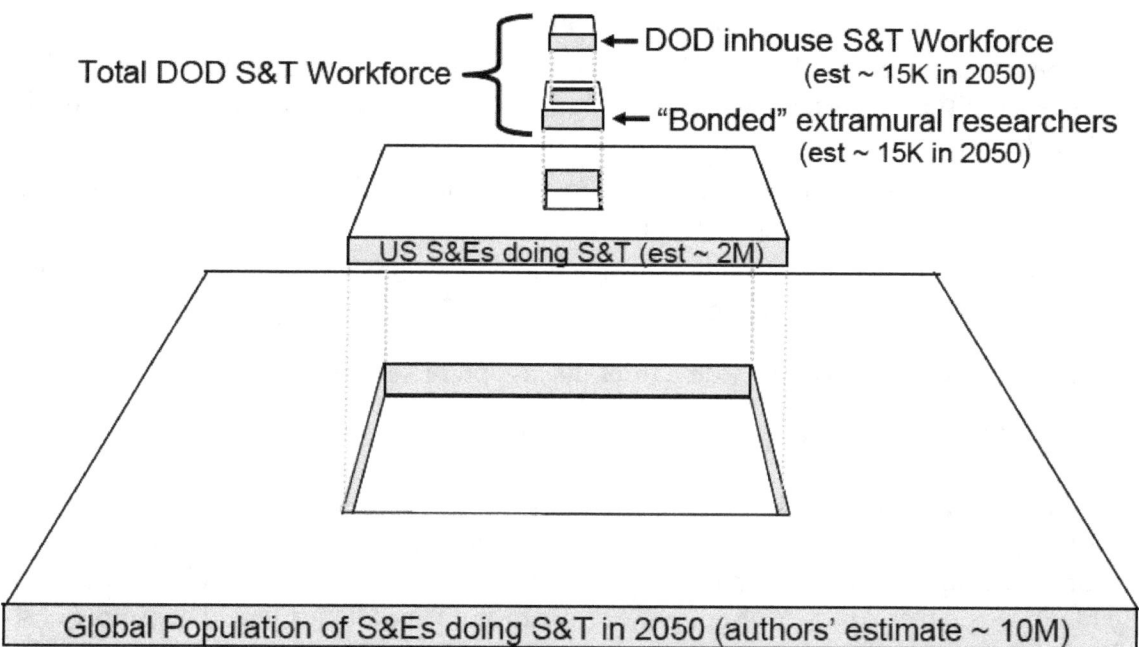

Figure 6. Visualization of Global S&T Workforce in Layers for 2050 According to Authors' Estimates of Population in the Four Components or Layers

This visualization m akes the challenges quite a pparent. First, the number of m embers in each layer decreases dram atically from Layer 1 to La yer 4, as indicated by the area of each layer (about 10 million globally in Layer 1 to about 15,000 in DOD in Layer 4). Clearly, our largest window on the world of global S& T is our national S&T workforce. This workforce, however, spends most of its tim e not thinking about nati onal defense. Gaining ac cess to defense-relevant knowledge resident in the U.S. national S&T workforce will be a significant challenge for DOD.

It is instructive to briefly compare the situation depicted in Figure 6 to the era when the United States dominated the global S&T enterprise. In th is era, the United States was about one-half of Layer 1 and the DOD in-house S&T workforce was a bout one-tenth of the U.S. S&T workforce. In that era, a si ngle DOD federal S &T worker was, on average, outnumbered globally about 20:1. By 2050, that ratio will have increased to   about 800:1. The numerical consequences are clear. In the future, each DOD S&E will have significantly more colleagues to track globally and domestically to maintain awareness of S&T developments. This scenario has major implications in regard to the DOD S&T workforce. It certain ly suggests that the quality of each DOD federal S&E must be high and that effo rts must be undertaken to im prove connectivity between the DOD S&T workforce and national S&T workforce.

The Federal Government only has a few avenues available to improve this situation:

- Within budgetary constraints, properly size and empower the DOD in-house S&T workforce so it represents the best of the national S&T workforce in S&T areas of long-term interest to DOD.

- Improve coordination and dependencies among the DOD in-house S&T workforce and the in-house S&T workforce employed by other government agencies.

- Within budgetary constraints, manage a contract research and technology portfolio that provides stable funding and connects a segment of the national S&T community to DOD.

- In the education curriculum for S&Es, include studies that focus on the civic responsibilities S&Es have for the national well-being, including matters of national defense.

- Make optimum use of TF and data mining techniques by the DOD S&T workforce in the quest for authoritative global S&T awareness.

- Balance the competition regarding the need to control S&T information and the need for open exchange of S&T knowledge.

The following sub-sections will provide a brief discussion of several of the above topics.

4.1 The DOD S&T Workforce

Because the national defense is among the h ighest of the Government's responsibilities, it is essential that the S&T workforce supporting that mission represent the nation's best S&Es. In this regard, the rule s for participation in the S&T co mmunity are well understood and immutable: one is either a play er or not player—there is no middle ground. The people best able to maintain awareness o f progress in S&T are those contributing to that S&T progress. Such individuals must form the core of the DOD S&T workforce. A subset of this workf orce should also have an awareness of potential m ilitary applications, which inv olves familiarity with national security issues and general m ilitary CONOPS or CONUSE to understand ho w something new might be useful or could become disruptive.

An essential component of the DOD S&T workfor ce is the federal or in -house component. This workforce plays an im portant role in advanc ing defense-related S&T, connecting DOD to the larger S&T community, identifying im portant S&T developm ents, and advocating the exploitation of those developments. To most effectively accomplish these responsibilities, the in-house workforce m ust be widely recognized as card-carrying m embers of the larger S&T community. This recognition is param ount to a chieving awareness because the total num ber of DOD in-house S&T researchers will not increase in proportion to the growth of the global S&T enterprise. DOD must be able to hire from among the nation's be st S&Es. Doing so may require hiring, employment, and com pensation initiatives, as well as organizational imperatives to prioritize and support S&T awaren ess activities. The U.S. Governm ent has em ployed similar recruiting initiatives in the past to meet shortages in certain fields, but these must focus on and be tailored to DOD's in-house S&T workforce and the ac tivities associated with awareness. In that sense, it is not prim arily a financial com pensation issue—although cer tain thresholds are probably required. It is really about creating a worki ng environment that attracts the bes t and then fosters their achievem ent and global re cognition. Being a DOD in-house S&E should be widely seen as performing important, challenging, and satisfying work. Much of the DOD's in-house S&T is performed by the three service corporate laboratories (Air Force Research Laboratory [AFRL], Army Research Laboratory [ARL], Naval Research Laboratory [NRL]). These laboratories should have a special responsibility for maintaining S&T awareness. Therefore, there is great merit in focusing the in-house component of the S&T workforce in the three DOD corporate laboratories so the proper culture can be nourished and critical masses of card-carrying members in many disciplines can be achieved.

While it is unnecessary to grow the in-house S&T workforce in pace with the international S&T workforce, it is nevertheless advisable (within budgetary constraints) to grow this workforce. Coffey has pointed out that the defense workforce has not kept pace with the national S&T workforce regarding developing areas of S&T. [23] This is worrisome. It will be necessary for the in-house workforce to maintain expertise in traditional defense areas. However, entirely new areas of S&T will emerge by 2050 that will be important to DOD. It is essential that the DOD in-house S&T workforce have expertise in these emerging areas. It is recommended that the DOD in-house S&T workforce be doubled over the next 30 years to track these developments while simultaneously maintaining expertise in the traditional areas. This growth should be used to establish expertise in the emerging areas. The resulting 2050 in-house S&T workforce of about 30,000 S&Es would then represent about 25 percent of the expected total DOD S&E workforce in 2050.

External researchers from academia and industry must complement the in-house component of the DOD S&T workforce. These individuals must also be card-carrying members of the global S&T community. They must be managed and funded by federal program managers who themselves are (or were) card-carrying members of the S&T community and who possess a keen awareness of military needs. An effort should be made to establish a bond between external researchers and DOD and its S&T aims to extend the reach and effectiveness of the DOD S&T awareness network. Indeed, while external researchers are likely to be funded by other sources as well, their bond with DOD will yield better awareness functions for DOD. These bonds become a means for attracting highly qualified individuals to DOD employment. Furthermore, external researchers can enable DOD to better mobilize sectors of the larger S&T community in times of acute and pressing needs and give DOD a knowledge base suitable to prepare for the war after next.

Although DOD cannot and should not guarantee funding to any component (in-house or not) of the S&T community, it should embrace the concept of a DOD S&T community and work to facilitate interaction and cooperation within this community. It is the impression of the authors that such a relationship existed between a subset of the national S&T community and DOD for several decades after WWII. This post-WWII arrangement, as typified by the early Office of Naval Research (ONR), was very effective in advancing DOD S&T interests. Unfortunately, the early bonds that were established were substantially weakened as DOD became a lesser player in the national S&T program and as the concept of a DOD S&T community slipped away. It would be helpful, in addressing the globalization of S&T, to create a 21st century analog of the post-WWII DOD S&T community.

---

[23] NDU Defense & Technology Paper Number 49 (2008).

An impediment to accomplishing the above is a tendency that has emerged over the past two decades to force federal agencies to compete in the marketplace along with all other performers for all or much of their funding support. This competition is usually counterproductive because, among other things, it compromises the "honest broker" function of the federal S&Es and causes the nongovernment workforce to view the in-house workforce as competitors, thereby impeding knowledge sharing within the larger S&T community. This is not conducive to building necessary partnerships and bonding. The tendency is also puzzling in that the in-house government workforce is one of the few areas remaining where government agencies can make truly strategic decisions, such as maintaining long-term competence in particular areas and undertaking high-risk projects that are not ready for competitive procurements.

Within the Government, the most successful S&T performing organizations are those where the sponsors enter a long-range strategic relationship with the federal S&T performing organization and hold that organization to a high standard of performance. Federal S&T performing organizations that operate according to this approach are among the most productive in the Government and compare very favorably with nonfederal S&T performing organizations when the standard metrics used to judge such organizations are applied. The federal S&T performing organizations must value these metrics and make them a priority while providing state-of-the-art supporting labs and facilities to these ends. In this case, the S&Es within the organization are empowered to become and remain card-carrying members of the S&T community through a culture that rewards the researchers for focusing on S&T and its frontiers. To the extent that these individuals are among the leaders and best researchers in the field, the job of awareness is easier and will naturally reach further. It is simply a matter of how many people will call the DOD S&E with new findings versus the DOD individual contacting an increasing number of dispersed researchers to learn of new results. Having a workforce of uniformly high-quality S&Es will also increase rates of discovery and invention, which can only further DOD's awareness. It ultimately comes down to interactions among subject matter experts on a global scale. One must encourage these interactions, learn from them, and plan from them. This was achieved when the United States dominated global S&T and national security was a primary motivation and major sponsor. In this regard, it is worth noting again that today's technical dominance by the U.S. military was established in that timeframe and in that context.

Even if the DOD S&T community is reinvigorated, the problems confronting DOD as a result of S&T globalization will be formidable. DOD will not have the fiscal resources to buy its way out of the problem. Some nonmonetary means must be found to motivate the national S&T community to accept some responsibility for keeping DOD aware of global S&T developments that could have significant national defense implications. In this regard, most of the U.S. national S&T workforce of 2050 is yet to be educated. Perhaps the education system is where the motivation should be developed as part of efforts to imbue students with an understanding of their profession's civic responsibilities. A concern for the health of national defense should be among those civic responsibilities. It may be that an acceptance of this civic responsibility among the national S&T community is essential to solving the DOD S&T awareness problem.

4.2 DOD S&T Programmatics and Portfolio Methods

It seems clear that the size of the DOD S&T program will suffer a continuous decline between now and 2050 relative to the global S&T program. This scenario requires a programmatic strategy regarding the details of the DOD S &T program. A research portfolio with fewer resources is typically enhanced by focusing attention in certain areas perceived to be of long-term future DOD i mpact.[24] Important areas will include those that have a lways been o f particular, perhaps unique, DOD interest, as well as em ergent S&T areas and so-called "convergent" areas at the intersection of more traditional S&T subject areas.[25] It is probably wise to err on the side of inclusion rather than to miss an area that will grow in importance. It will be necessary to distribute the mix of in-house and external brains cells in some optimal way across a map of the S&T portfolio of investm ents. The need to focus limited resources poses a dilemma to achieving broad awareness. Given lim ited resources, one can m ake an argum ent for employing contract research or limited investments in areas where the long-term interest is yet to be established and reserving in-house people and programs for m ore persistent topics and emerging areas where DOD i mportance has been esta blished. One must appr oach this carefully, however, or it will re sult in an in-house S&T workforce that is not well m atched to em erging S&T that will become essential to DOD. All of this must occur in the context of the larger U.S. S&T portfolio. Means must be devised to better link those other activities to the national security mission and DOD brain cells. The judg ments and balances needed to accom plish the above will be especially challenging and should be subject to careful scrutiny and evaluation.

DOD solicitations for new research should be as broad as possible to cr eate an inventory of proposals that is larger than will be funded. The external funding vehicle of choice should retain as much simplicity and flexibility in execution as possible. The proposal process should be highly interactive with constant feedback to foster convergence of ideas with the highest priority DOD areas of interest. Further, solicitations should focus on S& T content with only passing reference to military applications and use. These details can come later when the ideas are on the table. Broad solicitations allow DOD to m onitor new ideas and potential m ovements in frontiers of knowledge and applications beyond those th at DOD ultim ately funds. In this way, the portfolio will evolve.

If DOD would like to see proposals and ideas from the global enterprise to improve awareness, it should engage globally by presen ce and funding, or the global ente rprise will not subm it such proposals in the long run. All DOD S&T agencies m aintain a global presence and becom e involved in joint programs with other nations and agencies. This has been productive in the past, but the need for these activitie s will increase and should be an explicit component of the DOD S&T portfolio—a norm rather than a marginal activity. The desired norm is discussed briefly in the next section.

4.3 Interagency and International Activities

Even when the above is accom plished, it seems very likely that an expansion of the com bined DOD federal S&T workforce plus the DOD-bonded m embers of the larger S&T workforce will be desirable as a result of the m assive growth expected in global S&T. This expansion could be accomplished by encouraging and expanding joint S& T programs involving the federal S&Es of other U.S. agencies. While joint programs may involve some compromise in the specific goals of the original DOD investm ent, the gains from broader awareness and in creased S&T outputs are

---

[24] *Quadrennial Defense Review* (QDR 2010), pp. 94–95; see http://www.defense.gov/qdr/.
[25] *2025 Technology Intersections Study*, MITRE Technical Report, September 2007.

likely to outweigh those reductions.[26, 27, 28, 29] The 2050 context will likely benefit from far more such joint programs. While their primary aim is improved global S&T awareness, joint programs with other U.S. agencies will also enable DOD to be in a position to bette r mobilize broader sectors of the S&T community in times of special needs.

Similar comments apply regarding DOD and U.S. agency overseas S&T presence. An increased global presence seems intuitively wise for 2030 and beyond, but an effe ctive one may require careful study and adaptation. The character of the current activities may need to change in view of the relative shift in the global U.S. and DOD positions from 30 years ago (when most of these global outposts were established) to the context of 2030 and beyond.

4.4 Technology Forecasting and Data Mining

The ability to effectively manage the DOD S&T investment portfolio in the face of much larger U.S. and global inves tments will require access to techno logy forecasts and tech nology data mining of worldwide S &T activities. Also, with a decreasing DOD sh are of the global count of S&Es doing S&T, the DOD S&Es can enhance their individual awareness by similar access. The specific tools for each purpose—portfolio m anagement or awareness—may differ slightly but must be congruent and widely available to the users. We briefly discuss these tools, their features, and potential uses in this section and provide further details in Appendix A.

The activities termed "technology foresight" have largely sought to forecast the direction of S&T together with econom ic forces th at enable new develo pments to reach the commercial marketplace. In this sen se, technology foresight becomes a futures forecast of particular in terest to economic investors and policym akers. Often, it requires predicting one or m ore innovations from the advancing frontiers of science an d/or technology, as w ell as the necessary manufacturing methods. It m ay include concepts for novel m arketing strategies to achieve widespread successful implem entation in society. Of course, it is an inexact forecast for both technical and market reasons. For the purposes of DOD S&T awareness, we are less interested in forecasting arrival in the marketplace than in tracking movements at the frontiers of S&T and then rapidly assessing their poten tial impact on national security—either in term s of ne w opportunities or new threats—whether or not comme rcial market potential exists. It is also important for DOD to understand market forces th at could yield comm odity pricing for needed capabilities, as well as new th reats that m ight stem from the inte rsections of inexpensive widespread availability of a technology and result in unintended consequences.

However, DOD's need for global awareness of knowledge frontier m ovements is much m ore detailed and nuanced than the customary products of futures forecasting. Scientific frontiers can be moving in im portant and subtle ways well in advance (e.g., decades) of rapid progress an d

[26] *International Science and Engineering Partnerships: A Priority for U.S. Foreign Policy and Our Nation's Innovation Enterprise* (NSB-08-4, 2008).

[27] *Best Practices for Increasing the Impact of Research Investments*, Report by Ocean Research Advisory Panel of National Oceanographic Partnership Program (July 2007).

[28] *Army R&D Collaboration and the Role of Globalization in Research,* J. Lyons, NDU Defense & Technology Paper Number 51 (2008).

[29] *Globalization of Science and Engineering Research—A Companion to S&E Indicators 2010*, National Science Board (NSB) (2010); see http://www.nsf.gov/nsb/publications/index.jsp.

wide recognition.[30, 31] This is the prospecting phase of S&T that can lead to later, more evident, and rapid mining activities[32] and is the primary reason for having subject matter experts in DOD who are among the global S&Es working at the frontiers. These experts are properly placed and equipped to interpret the early events.

A subset of futures or technology foresight of more interest to DOD is often termed "technology forecasting," or TF, as described further in Appendix A. The aim of TF is to shape the DOD S&T investment portfolio over time. We briefly consider this portfolio in three main parts:

1. S&T areas of enduring and perhaps unique importance to DOD
2. Subjects identified as emerging S&T of potential and special relevance to DOD
3. Future investment topics of high potential for DOD relevance and little investment by others, as yet, where knowledge frontiers could move quickly.

In all cases, DOD must understand which S&T areas other sponsors may cover *and* make results available to DOD so its investments can be more focused to meet DOD needs. DOD must arrange its S&T portfolio to meet all three parts for investing S&T funds as described earlier. Doing so requires a strategy for employing the DOD S&T workforce. The various components of that workforce will have different focuses regarding the three parts of the portfolio. For example, the in-house component has a responsibility across all three parts. There are about 15,000 in-house S&Es funded by S&T dollars. Of these, about 5,000 work in the three corporate laboratories. Because the corporate laboratories are best positioned to remain current on emerging S&T, it is logical that they should have a special focus on parts (2) and (3). The remaining 10,000 in-house S&Es funded by S&T dollars, along with the industrial component, should focus on parts (1) and (2). The academic component should focus on (3). This is not to imply that the various components of the DOD S&T workforce should be restricted in the areas where they can contribute. Good ideas should be permitted to arise from any component of the workforce. Each component should, however, have a clear focus of responsibility and activity. DOD must optimally deploy its finite number of S&E assets to balance the competing demands of maintaining DOD S&T output and broad awareness. Only in this way can DOD arrange its portfolio to meet all nine of the purposes for investing S&T described earlier. Global S&T awareness is the key to success and can be aided through data mining, as discussed in the following paragraphs.

Performing S&T data mining to understand the frontiers of knowledge is not fundamentally new. If a literature review at the beginning of a research project is considered a form of technical data mining, as it should be, then researchers have been data mining since the dawn of scientific inquiry. It is useful to briefly describe the timeless process of literature review to see important elements and limitations. A review starts with a query of a particular topic within, for example, the scientific literature. From this, a number of potentially relevant papers emerge that the researcher carefully examines for content and for the list of referenced authors with related work. Most researchers then pursue other closely related topics suggested by paper's content and search for other publications and presentations by the cited authors of interest hoping for more recent and relevant "hits" on the subject(s) of interest. From this activity, new ideas and new

---

[30] NDU Defense & Technology Paper Number 17 (2005).
[31] *Science and Engineering Indicators 2010*, NSB (2010); see http://www.nsf.gov/nsb/publications/index.jsp.
[32] NDU Defense & Technology Paper Number 17 (2005).

collaborations can arise. New collaborations are critical to the researcher because they can y ield more timely awareness of advances by others s tarting even at the idea formulation stage. Such collaboration is only likely when the DOD S &E is viewed as a valuable peer to the other researcher (i.e., card-carrying). It is also worth noting this method's limits, which stem from language barriers of two ki nds: (1) the result of an actual shift in the wr itten language of the publication and (2) the missed detection of a relevant development in another field because it has evolved with another field-specific term inology and is unrecognizable from the seeker's disciplinary experience. In this case, em erging semantic methods show great prom ise provided the first disciplinary cross-over term inologies and concepts have occurred in the database searched.

What has changed in traditional literature review as a technical data m ining method is the means to search for information and the breadth of databases available to search. The co mbination of TF with technical data mining has been a very active area for more than a decade with the advent of the "inf ormation age" in a ve ry competitive global marketplace. A num ber of capable products exist. Several are described in de tail in two National R esearch Council (NRC) reports.[33, 34] We assume these tools will continue to im prove and will be s uccessful as long as they are widely available to the DOD S&E rese arch community engaged in the S&T awareness function. They m ust also be easy to use and inte rpret. Efforts toward the latter end are well underway. In a sim ple sense, the data m ining methods can reveal clusters and trends of activity by subject and/or by the researchers and institutions involved. Linkages between clusters and the generation of new clusters can be particularly enlightening. We note that searches of venture capital databases are now available and might have spotted sooner the unanticipated outcomes of the 25-year Army S&T Technology Forecast discussed in Appendix A.

Often, data m ining results are synt hesized into a diagram in the form of a network-like sketch. We give an exam ple of such an output in Figure 7. The sketch shown was done by Zhu and Porter[35] in 2001 to m ap global institutional involvement in nanotechnology. From such a m ap, based on analysis of text from published papers and reports, it is possibl e to indic ate the key subjects associated with each institution. In this map, the size of each node represents th e publishing activity on these sub jects at those in stitutions, and the indi cated links show the strongest correlations between the research activities of the various institutions. The rem aining notations (e.g., the ellipse) are used to give a three-dimensional perspective to the layout of the various subjects and linkages. More elaborate network maps are, of course, possible.

---

[33] *Persistent Forecasting of Disruptive Technologies*, NRC Report (2009); see http://www.nap.edu/catalog.php?record_id=12557.

[34] *Persistent Forecasting of Disruptive Technologies: Report 2*, NRC Report (2010); see http://www.nap.edu/catalog.php?record_id=12557.

[35] "Automated extraction and visualization of information for technological intelligence and forecasting," D. Zhu and A.L. Porter, *Technical Forecasting and Social Change* 69 (2002), pp. 495–506.

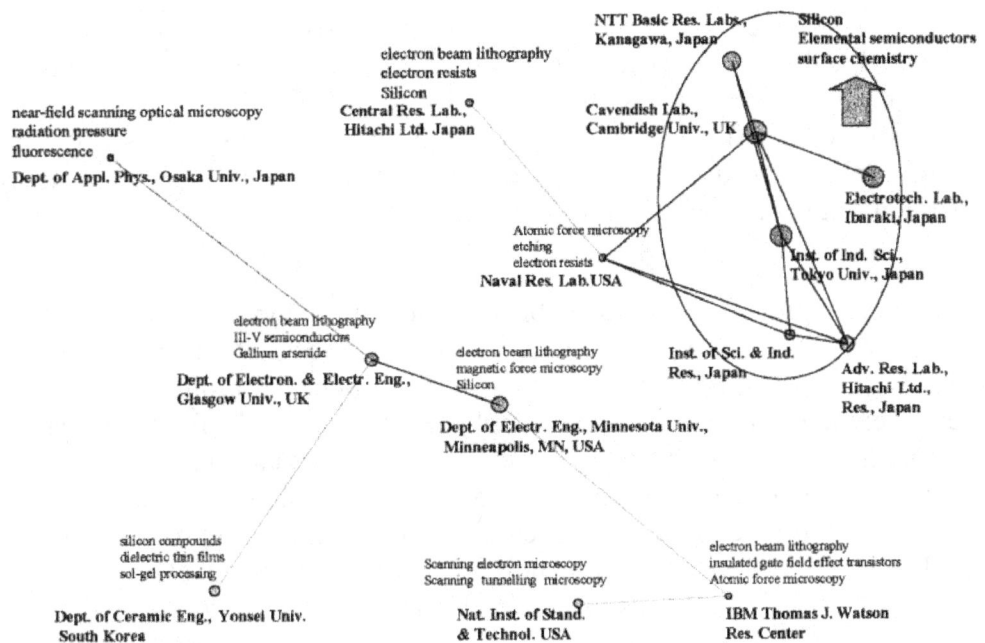

Figure 7. A Schematic of Results for Data Mining by Zhu and Porter[36]

It is clear that waiting for the literature to emerge is too late. For S&T awareness at the frontiers of new knowledge, the challenges become the timeliness and breadth of accessible databases rather than the tools themselves. We focus on these considerations in the next few paragraphs.

The timeline for a research effort extends over many years, and the presentation or publication of results occurs several years after the initial ideas are formed and the research begins. Formal publication may be preceded somewhat by availability of early results on the Internet, particularly the so-called "Deep Web."[37] Absent immediate awareness from collaboration by DOD S&Es with others, the next best opportunity is at the research proposal stage, where insight can be gained several years before new ideas and initial results may be recorded elsewhere. Obtaining such insights becomes possible via proposal databases through technical data mining tools and activities. The breadth of access should be across all U.S. S&T agencies, at the least, and internationally to the extent possible.

It becomes apparent that the most desirable database to search at this stage is the entire record of both funded and declined proposals. Indeed, this is the information base held by experienced DOD S&T program managers and by any DOD researcher who is engaged in the proposal review process as is expected for a card-carrying member of the community. Typically, this information base is broader than just the set of proposals to DOD because participants are individuals who are frequently involved in reviews for other sponsors both across and beyond the United States. Further insight into emerging S&T developments comes from automation and technical data mining tools applied to the entire proposal database. This search introduces a very significant issue of intellectual property rights and can create conflicts of interest. Since the aim

---

[36] Ibid.

[37] Content on the Internet that is not found in most search engine results because it is stored in a database rather than on HTML pages. Viewing such content is accomplished by going to the Web site's search page and typing in specific queries. LexiBot was the first search engine to make individual queries to each searchable database that it finds. Also known as the "invisible Web."

is basically a m apping of cluste rs and linkages of S&T activit ies, including the nam es of researchers engaged in them, one actually does not need the tec hnical concepts embodied in the proposal at first, on ly the keywo rds, the aut hor(s), the affiliation (s), and m aybe the cited references. This would be sufficient for th e data m ining mapping and should not involve proprietary information. The use of this inform ation, particularly if the concepts are to be explored as a result of the m apping, will have to be controlled to limit access to only authorized users (e.g., For Official Use Only).

This broad search of all proposals would enable early identification of emerging areas, as well as the beginnings of new disciplinary "cross-over " or convergence areas, even before any funding. These insights would provide new se mantic links between disciplines m uch sooner than previously possible; these links could then be applied to the full set of funded research databases *before* the new areas emerge in traditional literature sources. A data mining map that draws upon this data is also an excellent means to track progress and deviations from a technology forecast as time evolves and can be used to solicit structured inputs from subject matter expert surveys in, for example, the Delphi m ethod (see Appendix A). The insights can also be used to shape solicitations for new research proposals.

Lastly, the ability for scanning the h orizon for S&T activities will be an incre asing challenge to many U.S. agencies an d missions because of the global growth of the en terprise and staffing constraints in m any agencies and federal functions. [38] This challenge encourages a "whole o f government" approach to technology data mining activity.

4.5 Control of Information

A competition between the need to control S&T information and th e need for open S&T communications must be managed in the transition to 2050. This competition occurs at all levels. For example, basic res earch is generally char acterized as a culture o f open communication. However, researchers are reluctan t to release the results of thei r research until they are in a position to get credit (usua lly through a publication). At the sa me time, they also w ant to know what other researchers in their fi eld are doing so they can remain current, benefit from the work of others, and enhance the p rospects of th eir research. To acco mplish these objectives, researchers develop a strategy that involves asp ects of both an open and a closed system . The situation that confronts U.S. S&T (civilian and military) is very similar.

S&T is recognized as a key enabler for both m ilitary and economic competitiveness. As a result, there is a legitim ate interest in co ntrolling access in are as where one has a com petitive advantage. This interest is usually addresse d through security classi fication controls or proprietary information controls. On the other hand, one wants to gain information regarding the competitive advantages of others. The latter interest is addressed through a variety of approaches, such as m ilitary espionage, industrial espionage, and routine interaction am ong the S&Es of the various com petitors. As with all such undertakings, the quality of the resu lts is determined by the expe rtise of the various players and an understa nding of what is gained and what is lost by playing (or not playing).

---

[38] *The State of Scientific and Technical and Weapons Intelligence Analysis: Survey Findings*, K. Hawker et al., MITRE Report—For Official Use Only (2008) (or see *Studies in Intelligence* Vol. 52, No. 2 [2008]).

S&T is moving from a situation where the United States was the principal generator of scientific knowledge to one where m ost scientific knowledge will emerge from economies other than the United States. It will becom e increasingly difficult for the United States to mainta in an authoritative awareness of the state of S&T. A related development is that private companies are becoming increasingly global in character and will become difficult to associate with a particular nation. However, these entities may gain a better global view of S&T than any one nation, which could either help or hinder the matter of maintaining an authoritative awareness of world S&T.

We are also seeing the em ergence of global S&T initiatives (the human genome project, climate change, Large Hadron Collide r, ITER, etc.). F urthermore, the im pact of modern IT and the evolution of a new communication in frastructure greatly facilitate global collaboration in S&T (e.g., about one-third of U.S. sc ientific publications are now published in collaboration with other countries). These are profound changes, and we are only at the beginning of the globalization of S&T. Keeping up with these c hanges will be challeng ing for the United Sta tes and especially for DOD. The United States (inclu ding DOD) will need to balance the benefits of controlling S&T information to preserve military and economic advantages with the benefits of open exchange of S&T information to exploit global knowledge for U.S. intere sts. This will be a nontrivial and dyna mic calculus. A s with m ost balancing acts, leaning too far in one direction will have a negative outcome.

Most nations have in place pro cedures to prevent the release of information that could endanger their national or economic security. In the area of national security, a major means of control has been the security classification system. The U.S. security class ification system is very well developed and has three basic classification leve ls: Top Secret, Secret, and Confidential. These categories are defined in Executiv e Order 12356. The security cla ssification system applies to S&T information that has been derived fully or in part from federal funding. In general, it has been national policy that funda mental research (as defined in National Security Decision Directive 189) funded by federal funds should, to the m aximum extent possible, be unrestricted. There are, of course, procedures for classifying a fundamental research program in the event it is deemed necessary (e.g., Executive Orders 12958 and 13292).

A fourth category, referred to as Sensitiv e But Unclassified (SBU), is of ten invoked. This category has no statuary definition and has been the subject of cons iderable debate over the past years. This debate becam e especially loud after 9/11 when steps were taken to attempt to deny terrorists information that m ight be unclass ified but help ful to their cause. Unlike class ified information, which generally has well-defined cla ssification criteria, the SBU category is often ambiguous and characterized by subjective decisi onmaking. In 2006 the Congressional Research Service prepared a report on th is matter and the various positions and options a ssociated with it.[39] In 2008 the White House issued a memorandum for the heads of executive departm ents and agencies. It instituted the "Controlled But Unclass ified Information [CBU])" category that incorporated most of what was generally placed into the SBU category.[40] This memorandum was followed by the estab lishment of the Inte ragency Taskforce on Controlled but Unclassified

---

[39] See http://www.law.umaryland.edu/marshall/crsreports/crsdocuments/RL33303_02152006.pdf.
[40] See http://georgewbush-whitehouse.archives.gov/news/releases/2008/05/20080509-6 html.

Information. On August 25, 2009, the Task Force reported its findings. [41] The impact of the findings was not clear at the time this paper was published.

The private sector has sim ilar procedures for preventing the undesi red release of S&T information. These procedures generally fall under the categories of trade secrets and proprietary information. Both the national security and priv ate sector control syst ems involve a conscious tradeoff among the costs and benefits of control. For example, there are real and substantial direct costs associated with controlling inform ation. One must put in p lace an infrastructure for determining and codifying the rules for cont rol and ens uring the rules are res pected and enforced. De facto "opp ortunity costs" are also associated with the control of S&T infor mation, such as the loss of S& T progress and gains that would have resulted from the i nformation's release. These opportunity cost s must be weighed against th e benefits associated with information control, which include d elayed progress and increased costs faced by other nation s in the developm ent of arm aments and com peting products. A.S. Qui st has provided a m ore complete discussion of the benefits and risks associated with information control. [42]

In managing the transition to the global S&T en terprise of 2050, the balance am ong the risks, costs, and benefits of control ling information will also shift. Careful an d changing judgm ents will have to be m ade. It is cle ar that the choices for a world in which one party is do minant will differ from a world of a half-dozen or so equal players. It would seem unwise to take positions now that we would rue by 2050 when the United St ates will be looking for m ore exchanges and cooperation. Perhaps, as a cons equence of 9/11, the United Stat es has shifted—without careful analysis of risks, costs, and benefits—in a direction that will not serve us well in 2050. [43]

---

[41] See http://www.dhs.gov/xlibrary/assets/cui_task_force_rpt.pdf.
[42] See http://www.fas.org/sgp/library/quist2/index_html.
[43] *Beyond Fortress America: National Security Controls on Science and Technology in a Globalized World*, NRC Report (2009); see http://www.nap.edu/catalog.php?record_id=12557.

5. CONCLUSIONS AND RECOMMENDATIONS

The authors reached the following conclusions:

- By the middle of the 21st century, the U.S. share of the global S&T enterprise will decrease, and only a small fraction of U.S. S&Es will work on national security problems.

- The U.S. share of S&T productivity will decline from about 26 percent in 2005 to about 18 percent in 2050.

- The United States will remain one of the world's most significant contributors to scientific knowledge, and the U.S. S&T workforce should be large enough, relative to the world S&T workforce, to remain cognizant of S&T developments around the world. The DOD S&T workforce alone will not be large enough, and maintaining authoritative awareness of S&T globally will be among DOD's greatest challenges.

- Maintaining an authoritative awareness of S&T around the world will be essential if the United States is to remain economically and militarily competitive. The required awareness can be maintained only if the U.S. S&T workforce is a participant in the global S&T community. This is true for the DOD S&T workforce as well.

- The people best able to maintain authoritative awareness of progress in S&T are those contributing to that S&T progress. Such individuals must form the core of the DOD S&T workforce.

- It will be necessary to find means to tap the global S&T knowledge that the national U.S. S&T community will have. We must ensure the knowledge of global S&T held by the national U.S. S&T community is available to the military and that DOD has the internal capability to comprehend and exploit this knowledge through the DOD S&T workforce.

- TF and data mining will play increasingly important roles in maintaining global S&T awareness.

- In the transition to 2050 and beyond, DOD must wisely manage the legitimate competition between the need to control S&T information and the need for open S&T communications.

- Even if the DOD S&T community is reinvigorated as suggested in this paper, the problems confronting DOD as a result of S&T globalization will be formidable and beyond the ability of DOD alone. Some nonmonetary means must be found to motivate the national S&T community to accept some responsibility for keeping DOD aware of global S&T developments that could have significant national defense implications.

The authors offer the following recommendations:

- DOD should take the actions necessary to ensure the DOD S&T workforce is plugged into the national S&T community broadly (and to the extent possible into the global S&T community). To accomplish this, the DOD must build a DOD S&T workforce (including an in-house S&T workforce) with staff who are widely recognized as card-carrying members of the larger S&T community.

- The three primary DOD corporate research laboratories (AFRL, ARL, and NRL) should be assigned a special responsibility for maintaining authoritative awareness of and participating in emerging S&T believed to be of long-term potential importance to DOD.

- Coordination among DOD and U.S. agen cies engaged with the non-DOD S&T community must increase. The objective s hould be m ore joint programs among these agencies to help connect DOD with the larger S&T community and vice versa.

- The DOD S &T community should be utilized to authoritatively inte rpret the data that will emerge from the TF and data mining communities.

- Policies and procedures for infor mation control should be reevaluated to determ ine a strategic balance between the ri sks, costs, and b enefits of S&T information control in a 2050 context.

- DOD should work with various pr ofessional organizations and e ducational institutions to ensure that those receiv ing education related to participa tion in the future national S &T workforce consider the health of national defense to be among their civic responsibilities.

Appendix A. Technology Forecasting

We view technology foresight in three main forms:

1. Futures
2. Technology forecasting (TF)
3. Frontier movements in science and technology (S&T), including intersections or convergences.

In some sense, "futures" and "technology forecasting" are the more traditional concepts, and each depends on those listed below it. Therefore, both of the traditional activities ultimately depend on the third concept—estimating movements in the S&T knowledge frontiers in some relevant timeframe. DOD's need for global awareness of these movements of knowledge frontiers is much more detailed and nuanced than the traditional activities provide. Some advances may have no relevance to DOD, or advancement of a very particular feature may be particularly relevant and yield important clues about future capabilities. Also, scientific frontiers often move in important and subtle ways well in advance (e.g., decades) of rapid progress and wide recognition. This is the prospecting phase of S&T that can lead to later, more evident, and rapid mining activities.[44] Subject matter experts in DOD—who are among the global scientists and engineers (S&E) working at the frontiers—are essential for the third concept to succeed for DOD. These experts are properly placed and equipped to interpret the early events.

A useful definition of TF is given in the recent NRC study *Persistent Forecasting of Disruptive Technologies*.[45, 46] The definition expands on an earlier definition in Martino.[47] The NRC study takes TF to be "the prediction of the invention, timing, characteristics, dimensions, performance, or rate of diffusion of a machine, material, technique, or process serving some useful purpose." This interpretation also highlights that TF is largely about events *after* there has been movement in the frontiers of S&T knowledge when the "useful purpose" and even the nature (e.g., "machine," "material," or other) of the movement are more fully evident. The NRC study goes on to describe TF methodologies in four general categories:

1. Judgmental or intuitive methods
2. Extrapolation and trend analysis
3. Models
4. Scenarios and simulation.

Within these general categories are many individual methods, as described in the NRC report. For reference, Table A-1 is reproduced from the *Handbook of Technology Foresight*,[48] which lists the 33 common methods of TF in columns according to the degree of judgment or intuition (e.g., "qualitative" nature).

---

[44] NDU Defense & Technology Paper Number 17 (2005).

[45] *Persistent Forecasting of Disruptive Technologies* (2009); see http://www.nap.edu/catalog.php?record_id=12557.

[46] *Persistent Forecasting of Disruptive Technologies: Report 2* (2010); see http://www.nap.edu/catalog.php?record_id=12557.

[47] "Recent Developments in Technological Forecasting," J.P. Martino, *Climatic Change* 11, pp. 211–235 (1987).

[48] *The Handbook of Technology Foresight—Concepts and Practice,* L. Georghiou et al., Edward Elgar Publishing (2008).

Table A-1. Common Methods of TF

Qualitative	Quantitative	Semi-Quantitative
1. Backcasting	20. Benchmarking	26. Cross-Impact/Structural Analysis
2. Brainstorming	21. Bibliometrics	27. Delphi
3. Citizens Panels	22. Indicators/Time Series Analysis	28. Key/Critical Technologies
4. Conferences/Workshops	23. Modeling	29. Multi-Criteria Analysis
5. Essays/Scenario Writing	24. Patent Analysis	30. Polling/Voting
6. Expert Panels	25. Trend Extrapolation/ Impact Analysis	31. Quantitative Scenarios
7. Genius Forecasting		32. Roadmapping
8. Interviews		33. Stakeholder Analysis
9. Literature Review		
10. Morphological Analysis		
11. Relevance Trees/Logic Charts		
12. Role Playing/Acting		
13. Scanning		
14. Scenario Workshops		
15. Science Fictioning		
16. Simulation Gaming		
17. Surveys		
18. SWOT Analysis		
19. Weak Signals/Wildcards		

Most technology forecasts combine several of these methods in a particular study to achieve a balance among the strengths and weakness of a particular method. The TF studies can be also be characterized[49] as projecting from today in an orderly, mostly traceable, fashion or leaping to a future state where the path to that state is not well specified ("continuous" or "discontinuous"). The objective of the study can characterize it as "exploratory," wherein the aim is to consider the evolution of some technology, for example, and then decide what must be done to get there; or, the study can be termed "normative," wherein future opportunities or threats are assessed for a forecasted evolution in one or more technologies, for example. For DOD purposes, both exploratory and normative aspects are usually combined in a more or less continuous perspective on the future evolution. The typical timeframes for such studies in DOD fall into short term (within 5 years), medium term (5–10 years), and long term. The nature of TF also shifts with the timeframe, typically going from exploratory in the short term to normative in the long term. The occasional longer term or discontinuous look can mitigate risks and avoid surprises by subsequently placing a few "brain cells" in the areas identified as outliers in these results. The notion of monitoring outliers to a consensus view can be formalized using the Weak Signals/Wild Cards (#19) method.

The typical TF approach used in DOD is some combination of the following methods:

---

[49] *The Handbook of Technology Foresight—Concepts and Practice* (2008).

- Brainstorming (#2)
- Conferences/Workshops (#4)
- Expert Panels (#6)
- Key/Critical Technologies (#28).

The above methods are used togeth er with el ements of Weak Signals/W ild Cards (#19), Tren d Extrapolation/Impact Analysis (#25), or Roadm apping (#32) in varying de grees. Note these are mostly qualitative methods that rely largely on individual judgment and expert intuition.

There are numerous examples of such studies. We cite a report that examined the performance of an earlier long-term TF for Ar my S&T[50] and then m ade recommendations to improve the approach by considering convergences between S&T areas and use of Roadmapping (#32) and Stakeholder Analysis (#33) TF m ethods. The assessment of the earlier 25–30-year long-term forecast—made 15 years later—sho wed that ab out one-quarter of the topics (5) w ere right on target, another quarter (4) underestim ated the s ubsequent pace of progress, and ano ther quarter (4) lagged the forecast progress. Th ree forecasted topics never substantially advanced at all, and the impacts of four rapid advances were m issed related to the explosion of the Internet, IT, and wireless communication applications in the late 1990s. One could argue that correctly identifying about three-fourths of the topics, incorrectly estimating the pace fo r one-half of the topics, and getting a few wrong by inclusion and a few by omission is fairly good and probably in the nature of such fore casts. It sug gests that a revisit of any such long-term forecast every 5 y ears or so would be wise, but doing so more often may not be useful. Rather, techniques to track deviations from the forecast by co mparison to regular tec hnology data m ining efforts m ight be wise, as discussed earlier.

We also note the m issed areas were all stimulated by developments outside DOD and were more in the natu re of very rapid techn ology exploitation in the m arketplace with unanticipated consequences for national security. As DOD becomes a sm aller player in the global S&T enterprise, such m issed forecasts are increasin gly likely unless m ore robust TF m ethods are employed that involve broad-based participa tion of subject m atter experts beyond DOD and the United States and a m eans to independently and regularly monitor th e evolution of the global S&T against this forecast. We note that the Delp hi survey method (#27) originally developed by RAND in the 1960s [51] to m itigate issues of "groupthink" an d of sm all number biases in exper t panels is promising in the web-ba sed context of today b ecause it can readily scale to larg er and more disperse participation. More over, broad survey participation stimulates the S&T awareness function of the participants. To the extent that individual expert opinion iterations in the Delphi method are also considered in a "weak signal" sense to target technical data mining activities, the overall process can become more robust.

---

[50] *Improving the Army's Next Effort in Technology Forecasting*, J.W. Lyons, R. Chait, and S. Erchov (ed.), NDU Defense & Technology Paper Number 73 (2010).
[51] See http://is.njit.edu/pubs/delphibook/.